QASS
QUR'AN AND SCIENCE SERIES

ASTRONOMY

By Nasima Ali

Dedicated to

Copyright © 2015 to present, by Nasima Ali
All rights reserved. This book or any portion thereof
may not be reproduced or used in any manner whatsoever
without the express written permission of the publisher
except for the use of brief quotations in a book review.

Printed in the United States of America

ISBN 978-0-9967319-0-4

Boost The Mind, LLC
Riverdale, MD
www.boostthemind.com

Contents

Dear Reader 1

Creation of Planets from Stars 3

Life Created from Water 5

Six Days?!! New Arabic Words 7

Regulation of the Heavens 9

Lamp Vs. Light: The Exactness of the Arabic Language 11

What are Shooting Stars? 13

Seven Heavens 15

Stages of the Moon 17

Night and Day 19

The Benefits of the Stars 21

What Does Creation Say About God? 23

The Zodiac Stars 25

Night & Day: Fluid Motion 27

Dark Matter, Part 1 29

Dark Matter, Part 2 31

Navigation via the Stars 33

Structure of the Universe 2 35

Movement of the Sun: Sun Dial 37

You Are Now Being Served 39

Comets and Asteroids 41

- Structure of the Universe, Part 3 43
- Benefits of The Moon 45
- Magnetism: Spin Up and Down 47
- Creation of Solar System, Part 2 49
- Index 51

Name _____

Dear Reader

Assalamu alaikum wa rahmatullah.

Thank you for purchasing and using this workbook. Our goal is to take you on a journey of discovery where the Qur'an is our guide. Allah Subhana wa Ta'ala (SWT) mentions His wonderful creations in His Book, so in order to really appreciate Allah, we need to learn about these creations. Inshallah, this book will help you do that.

Please know that this book is not an official "tafsir" or interpretation of the ayaat. To see what the tafsir is, please consult a scholar, or refer to the works of historians such as Ibn Kathir, Tabari, and so on.

We are taking each ayat that is about the universe, and using it to remind ourselves about something that Allah SWT has gifted us with.

If this book makes you curious about what else is out there, you should follow your heart and find out! Allah's Creation is vast and we can never know all that there is. Learning about His Creation can even be an act of worship.

Let us begin our journey with Astronomy.

Vocabulary

Allah: This name is used for God in Arabic. Allah is a name that is neither male nor female. It literally means "the one who is worshipped."

SWT or Subhana Wa Ta'ala: This epithet follows the name Allah. It is a praise statement, that literally means "glorified is He and raised up above all."

SAWS or Salla Allahu alaihi wa Sallam: This praise statement was mandated by Allah SWT in the Quran, where we are commanded to "send [prayers of] peace and blessings on the prophet."

Qur'an: This spelling is used to reflect the correct pronunciation for the Book of Allah SWT sent to Muslims via the Prophet Muhammad SAWS.

Name _____

Creation of Planets from Stars

21:30. Have not those who disbelieve known that the heavens and the earth were joined together as one united piece, then we parted them? And we have made from water every living thing. Will they not then believe?

"The heavens" in this verse means the stars. Allah SWT informs us that the stars and planets used to be one entity. But the stars and the planets seem to be very different. Is it possible the planets are made out of the stars?

In the 18th century, some scientists had guessed that the material for forming planets came from dying stars[1], but only now have astronomers been able to find strong evidence to support this idea. Now they have images of stars dying (which is an event called a supernova) and these pictures show a cloud of dust coming out of the supernova. In the images after the supernova, we see the dust settling down into clumps orbiting another younger star. The round clumps seem to attract each other and join up to make a planet.

Finally, other scientists have looked for and found rare elements on Earth that could only have come from a dying star. So most scientists have accepted the hypothesis that stars were created first out of the Big Bang, and then 14 billion years later, the first planets were born out from the stars that started dying in supernovas.

According to this passage, what are two pieces of evidence that support the idea that "the heavens and the earth were one"? HW Bonus Activity: Search for the elements that are found on Earth that only could have come from stars. What are their periodic table symbols and numbers such as atomic mass?

---

[1] This widely accepted model, known as the nebular hypothesis, was first developed in the 18th century by Emanuel Swedenborg, Immanuel Kant, and Pierre-Simon Laplace. [Wikipedia, formation evolution Solar System.]

Name _____

Name _____

Life Created from Water

...And we have made from water every living thing. Will they not then believe? (21:30)

Let's follow the scientific method and prove that "living things are made of water?"

- What are some things you have already observed about living things and water?

- What can you do to prove "Are living things made of water"? List all ideas, and then we'll narrow it down. For test subjects, you may use bacteria, algae, seeds or plants.

- What kind of test can you do on plants, to find out if your educated guess is correct or not correct?

If time permits, do some reading, and find out if the test you needed has already been done. If so, what were the results, and what do you conclude from those results? Perhaps your teacher can give you time to do the test. If so, fill in your conclusions after the test is done.

Name _____

Name _____

Six Days?!! New Arabic Words

59. [Allah is He] Who created the heavens and the earth and all that is between them in six days [yaum]. Then He Istawâ (rose over) the Throne (in a manner that suits His Majesty). The Most Beneficent (Allâh)! ... (Q S25. Al Furqaan, ayat 59)

New Words: Yaum, Ayyam, Istawa

Yaum: You see this word translated as "a day of 24 hours", but did you know it has other meanings? The Arabs can use "*yaum*" to mean any period of time. So when Allah SWT talks about six "days", He does not always mean 24 hour periods. In fact, we don't know how long these **ayyam** are unless they are defined. So yaum, and its plural form ayyaam, are better translated as "periods of time", not "days".

Write a paragraph where you use the word "day" or "yaum", but it does **not** mean 24 hours.

Practice writing "yaum" in Arabic and its plural "ayyam"

يوم _____

أيام _____

BONUS: What is Istawa? in the translation, this means "rose over the Throne" and then is followed by "in a manner that suits His Majesty" in parenthesis. This is a demonstration of an important principle in our belief (our *Aqeedah*). Muslims believe that Allah SWT is not like anyone else, including humans. So when He rises, sits on His Throne, speaks, or does other things that we also do, we say that He does these actions in the most perfect way that is unlike the way we do it, and so it is "in a manner that suits His Majesty".

Name _____

8

Name _____

Regulation of the Heavens

3. Surely your Lord is Allah, Who created the heavens and the earth in six periods, and He is firm in power, regulating the affair, there is no intercessor except after His permission; this is Allah, your Lord, therefore serve Him; will you not then mind?

[Yunus, 10:3]

Generally in science, we love to organize the history of things. We try to find stages of development, whether it is how a human develops, or how the earth got to the way it is today. What do you think Allah SWT is talking about when He says the universe was created in "six periods"?

Name _____

Lamp Vs. Light: The Exactness of the Arabic Language

61. Blessed be He who has placed in the heaven big stars, and has placed therein a great lamp (sun), and a moon giving light. (Q S25. Al Furqaan, ayat 61)

Subhanallah, Arabic is a very visual and precise language. The way that Allah SWT uses Arabic in the Quran was so masterful that it would convince the pagan Arabs to give up their idol-worship and join Islam.

1. What is the difference between the light of the stars, the sun, and the moon? Name all differences that you can think of.

___

___

___

___

___

___

___

___

___

___

___

___

Other Questions to Discuss:
Name and describe the chemical reaction that produces sunlight.
What caused the moon to have shiny flat surfaces called "mare" or seas in Latin?
How far is the sun and how far are the stars from us?

Name _____

Name _____

What are Shooting Stars?

16. And indeed, We have put the big stars in the heaven and We beautified it for the beholders.

17. And We have guarded it (near heaven) from every outcast Shaitân (devil).

18. Except him (devil) that gains hearing by stealing; he is pursued by a clear flaming fire. (S.15. Al Hijr, ayat 16-18)

Facts about Meteors:

- A "meteor," or meteor shower, is composed of meteoroids (rocks and debris) floating in the Earth's atmosphere.
- They appear as a streak of light in the sky. That is the meteoroid burning up in our lower atmosphere, as earth's gravity finally pulls it down to crash on earth.
- They are often called "shooting stars."

Have you ever seen a meteor shower? Have you seen a shooting star? If not, try this at home. Find out about any special events going on in the sky tonight. Maybe there's a meteor shower. Observe the sky for 15 minutes. Now, describe what you saw.

_____
_____
_____
_____
_____
_____
_____

Name _____

Name _____

Seven Heavens

17. And indeed We have created above you seven heavens (one over the other), and We are never unaware of the creation. (S23. Al Mu'minuun, ayat 17)

There are many interpretations of the phrase "Seven Heavens." One possible explanation provided by scholars is that there are seven other worlds in the solar system, if you don't count Pluto. Pluto was removed from the list of planets in 2009. And we don't count Earth, because most verses say "As samawati wal Ard" ie. the heavens and the earth.

Inner Planets	Distance from Sun (Astronomical Units)	Outer Planets	Distance from Sun (Astronomical Units)
Mercury	0.39 AU	Jupiter	5.20
Venus	0.72	Saturn	9.58
Earth	1.00	Uranus	19.20
Mars	1.52	Neptune	30.05

⬅ Draw a diagram of the solar system here, using a straight line between the planets. Label the distances. (Tip: Try to use a ruler and use a scale for your picture, where 1 cm = 1 AU. You will need the whole length of this page.)

Name _____

16

Name _____

Stages of the Moon

5. It is He who made the sun a shining thing and the moon as a light and measured out its (their) stages, that You might know the number of years and the reckoning.

Allâh did not create this but in truth.

He explains the Ayât in detail for people who have knowledge. (S10. Yunus, ayat 5)

The 8 phases of the moon are categorized based on how much of the moon you can see. After reaching the Full phase, which usually occurs on the 15th of each lunar month, the moon will slowly decrease in visibility until it is time for the New phase to begin.

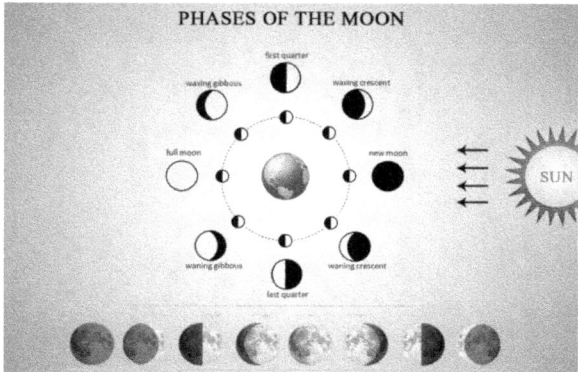

Pretend you are teaching a non Muslim friend who asks: How are the phases of the moon related to Islamic practices like fasting and festivals (Eid)? What about lunar eclipses in Islam?

_____
_____
_____
_____
_____
_____

Name _____

Name _____

Night and Day

62. And He it is who has put the night and the Day in succession, for such who desires to Remember or desires to show his gratitude. (S25. Al Furqaan ayat 62)

What if there were 23 hour in daylight, and the darkness of night was only an hour long? Or vice versa? How would these changes in the amount of day and night affect your activities and habits?

_____

_____

_____

_____

_____

_____

_____

_____

Name _____

Name _____

The Benefits of the Stars

12. And He has subjected to you the night and the day, the sun and the moon; and the stars are subjected by his Command. Surely, in this are proofs for people who understand. (S16. An Nahl, ayat 12)

What are the benefits of the stars?

_____

_____

_____

_____

_____

BONUS QUESTION:

What is astrology, and why does it conflict with Islamic values?

_____

_____

_____

_____

_____

_____

Name _____

Name _____

What Does Creation Say About God?

189. And Allah´s is the kingdom of the heavens and the earth, and Allah has power over all things. 190. Most surely in the creation of the heavens and the earth and the alternation of the night and the day there are signs for men who understand. [Al 'Imran,3:189-190]

When we look at how the universe was created, what do YOU think it tell us about the Creator?

Name _____

Name _____

The Zodiac Stars

96. He causes the dawn to break; and He has made the night for rest, and the sun and the moon for reckoning; this is an arrangement of the Mighty, the Knowing.

97. And He it is Who has made the stars for you that you might follow the right way thereby in the darkness of the land and the sea; truly We have made plain the communications for a people who know. [Al An'am, 6:96-97]

The stars can be used to navigate. It is important to know about this so we can thank Allah SWT for the stars. They are more than just beautiful points of light in the night sky. Because they appear in a similar place in the sky each night, they were used by ancient peoples to find their way when traveling. Clusters of stars that look like an image are known as constellations. The twelve most famous constellations are called the Zodiac. In the past, people were able to identify Zodiac constellations and guide themselves home by the stars.

In the diagram below, you can see which constellation is visible in May. You can also see how the constellations are imagined to be pictures, like a "connect the dots" puzzle. Four of these famous pictures have been left incomplete so that you can have some fun trying to connect the dots yourself.

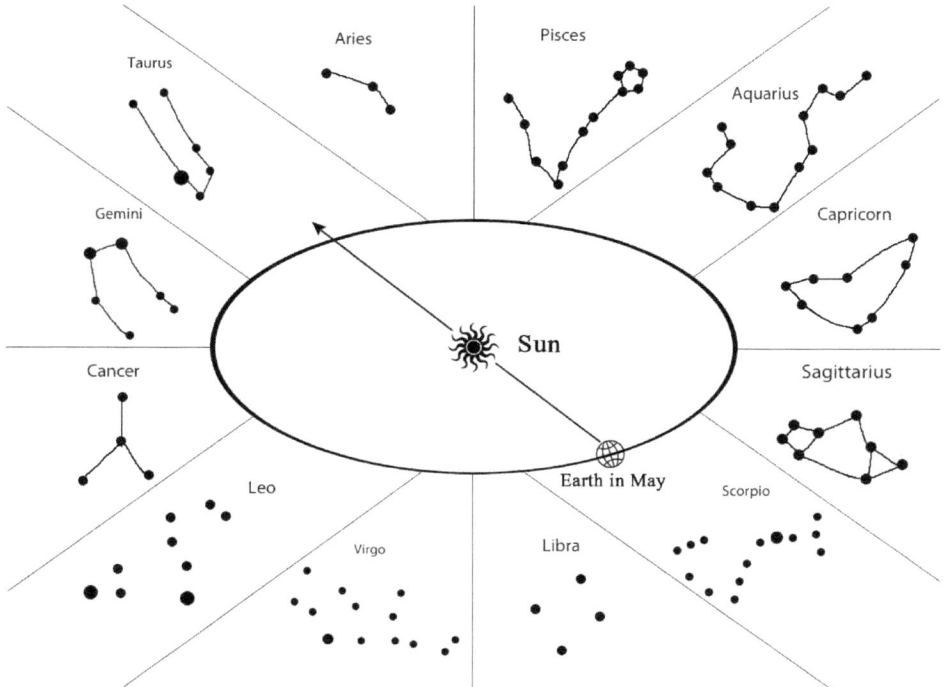

Name _____

Name _____

Night & Day: Fluid Motion

54. Surely your Lord is Allah, Who created the heavens and the earth in six periods of time, and He is firm in power; <u>He throws the veil of night over the day, which it pursues incessantly;</u> and (He created) the sun and the moon and the stars, made subservient by His command; surely His is the creation and the command; blessed is Allah, the Lord of the worlds.

[Al A'raf, 7:54]

I can't even begin to say how beautiful this is. "He throws the veil of night over the day, which it pursues incessantly". This is a form of comparison. Allah SWT is describing the night as a piece of cloth, moving fluidly over the earth in pursuit of day. When one object is described as another, that is called a metaphor.

Can you come up with some other metaphors for how Allah SWT runs the world? Or to make it easier, you can make similes, which are comparisons that use the words "like" or "as".

Name _____

Name _____

Dark Matter, Part 1

2. Allah is He Who raised the heavens without any pillars that you see, and He is firm in power and He made the sun and the moon subservient (to you); each one pursues its course to an appointed time; He regulates the affair, making clear the signs that you may be certain of meeting your Lord.

[Ar Ra'd, 13:2]

It will take many days to discuss the structure of the universe! To begin, let's talk about gravity and 'dark matter'. Gravity is something you've heard of and you've probably imagined that gravity has "reached out" from Earth and "grabbed" the moon. It seems to have pulled the moon into orbit around us. In the same way, gravity seems to also reach out and cause: the planets to come and stay in orbit around the Sun, the Sun to stay within the Milky Way galaxy, and this galaxy to orbit the center of the Local Cluster, along with 50 other galaxies.

So what is gravity? For now, let's call it a "force" even though it's more like a groove that rolls smaller objects towards bigger objects. The Earth is so much more massive than an apple, so the apple rolls toward the Earth. It is because earth has so many more atoms, i.e. lots of matter. So looking at matter has been a way of predicting how much gravity there is.

But two astronomers, Fred Zwicky and Vera Rubin, noticed that to make the amount of gravity or pulling force that we are measuring, there is not enough visible matter. The planets and galaxies are all moving as if something much more massive is pulling them than the amount of suns, planets, and stars that we can see. Therefore, there must be <u>something else we cannot see</u>, and that is being called "dark matter" and "dark energy". Visible matter accounts for only 15% of the matter needed for the gravity in the universe. So 100-15%=____ THIS is how much of the universe is actually Unseen!

For your Web Quest, have a look at these sites that further explain dark matter:

http://study.com/academy/lesson/dark-matter-lesson-for-kids.html

https://www.esa.int/esaKIDSen/SEM2RDW2EMH_OurUniverse_0.html

http://discoverykids.com/articles/dark-matter/

Name _____

Dark Matter, Part 2

33. And He has made subservient to you the sun and the moon pursuing their courses, and He has made subservient to you the night and the day. [Ibrahim:33]

What have you learned about gravity and dark matter from the previous page?

Name _____

Name _____

Navigation via the Stars

15. And He has cast great mountains in the earth lest it might be convulsed with you, and rivers and roads that you may go aright,

16. And landmarks; and by the stars they find the right way.

17. Is He then Who creates like him who does not create? Do you not then mind?

[An Nahl, 15-17]

An astrolabe is a handheld device resembling a compass that solves many problems related to the Earth being a sphere, in order to calculate qibla, prayer times, directions, time of day, rising and setting of Sun, moon, and stars, and location. And that's probably not all it could do! It was like a bunch of smartphone apps.

While the astrolabe itself was invented two thousand years ago, the universal astrolabe was invented by a Muslim by the name of Al-Zarqali (Arzachel). He redesigned the instrument so that it could be used anywhere on earth. The complex mathematics of the astrolabe were worked out by Al-Battani in 920 AD, and published in Europe with translation. A mechanical astrolabe was created by Abi Bakr Isfahani in 1235. It must have been like a mechanical clock.

Follow this link for further information on the astrolabe.

http://www.thepirateking.com/historical/astrolabe.htm

Name _____

Name _____

Structure of the Universe 2

65. Do you not see that Allah has made subservient to you whatsoever is in the earth and the ships running in the sea by His command? And He withholds the heaven from falling on the earth except with His permission; most surely Allah is Compassionate, Merciful to men.

[Al Hajj:65]

Allah says in this verse that He withholds the heavens from falling on us. It is because of his precise control of space that we are able to enjoy earth. Also, our planet is held in such a way that it is perfect for life as we know it. Let us list some of these ways that Allah protects us and positions us precisely where we need to be.

- We are protected from cosmic rays of high energy particles by the oxygen and nitrogen in Earth's atmosphere.
- Gamma rays and X-rays from outer space are absorbed by our atmosphere.
- UV radiation can't reach us because of the ozone layer of our atmosphere.
- Meteors are burned up in the atmosphere before they can reach our surface.
- Solar "wind" and plasma shoot towards us from the Sun. But Earth's magnetic field extends out into space and creates a "magnetosphere" that repels these particles and plasma. Protons and electrons get trapped in the Van Allen belts. They are the cause of the beautiful Northern Lights.

Use the space below to copy the image of the magnetosphere as your teacher instructs or from this site: https://www.windows2universe.org/earth/Magnetosphere/overview.html

Name _____

Name _____

Movement of the Sun: Sun Dial

45. Have you not considered (the work of) your Lord, how He extends the shade? And if He had pleased He would certainly have made it stationary; then We have made the sun an indication of it

[Al Furqan:45]

96. He causes the dawn to break; and He has made the night for rest, and the sun and the moon for reckoning; this is an arrangement of the Mighty, the Knowing.

[Al An'am, 6:96-97]

You may know that a sun dial is an easy way to track of time. This is one way in which the sun, and the shadows it makes, are in service to mankind. There are many websites devoted to how to make a simple sun dial, such as these two pages:

https://www.nwf.org/kids/family-fun/crafts/sundial.aspx

http://www.wikihow.com/Make-a-Sundial

The shadow of the central stick, or the centennial pole, will fall in the same place every day. So it is easy to know the hour, especially if you are familiar with the positions of the shadow at each hour. If not, the hours are marked on the dial around the stick so that you don't have to remember where the shadow will be at every hour.

The length and position of the shadow has been mentioned in various hadith to help us calculate daily prayer times. Can a traditional sun dial help us to do this? <u>Yes. You can modify your sun dial by marking off prayer times. Notice the length of the shadow, not just its position.</u> In the Islamic world, there are many example of elaborate sun dials used to help find prayer times and qibla, such as the one in this video:

What Do Sundials Have to Do With Islamic Prayer? Video on youtube.com

Name _____

Name _____

You Are Now Being Served

61. And if you ask them, Who created the heavens and the earth and <u>made the sun and the moon subservient</u>, they will certainly say, Allah. Whence are they then turned away?

[Al 'Ankabut:61]

29. Do you not see that Allah makes the night to enter into the day, and He makes the day to enter into the night, and <u>He has made the sun and the moon subservient (to you)</u>; each pursues its course till an appointed time; and that Allah is Aware of what you do?

[Luqman:29]

"Subservient" in Arabic is "Sakkara" وَسَخَّرَ ٱلشَّمْسَ وَٱلْقَمَرَ

What does it mean in English?

The prefix "sub" means _____.

And the root of the word is "serve". So to be subservient means ...?

Now can you list one or more ways that the sun serves us?

Name _____

Name _____

Comets and Asteroids

9. Do they not then consider what is before them and what is behind them of the heaven and the earth? If We please, We will make them disappear in the land or bring down upon them a portion from the heaven; most surely there is a sign in this for every servant turning (to Allah). (34:9)

Comets and asteroids both orbit the Sun. However, comets are known to have very large orbits that take them out of the solar system. Also, comets are made up of ices and dust. The ice could be frozen water, or it could be another compound, such as methane, ammonia, or carbon dioxide. When the comet comes near the Sun, the ices melt and a comet gas tail is formed. It is believed that comets were born when the solar system was first molded.

Asteroids are also as old as the Solar system. They tend to be found in one of two asteroid belts. The first Asteroid belt is between Mars and Jupiter. In this belt, there are a couple million asteroids (roughly), ranging in size from pebbles to 600 miles across. They are not termed planets because of their size, but they are made of rocks and metals. The other location where most are found is the Oort Cloud, which is located beyond Neptune. Pluto is in the Oort Cloud. There are trillions of objects in that cloud, and it extends to the end of the Solar System, about 20,000 AU from the Sun.

Create a Venn Diagram comparing and contrasting comets and asteroids.

Name _____

Structure of the Universe, Part 3

41. Surely Allah upholds the heavens and the earth lest they come to naught; and if they should come to naught, there Is none who can uphold them after Him; surely He is the Forbearing, the Forgiving. (35:41)

We are ready to discuss dark energy, now that we know about dark matter. Remember that Einstein said "$e=mc^2$". In other words, energy and matter are the same thing. If you take matter and get it moving at twice the speed of light, then it turns into energy.

We also calculated that only 15% of the universe is Seen or visible. The Unseen portion is about 85%. Given the above equation, astronomers decided that of the 85% that is unseen or "dark", only 30% of it is "dark matter". The rest of it is "dark energy".

Not too much is known about dark energy. But it is thought that dark energy is the energy that actually is pushing everything away from the origin of the Big Bang. It is acting opposite to the gravity produced by matter. The second thing that is known about dark energy is that it becomes stronger as distance increases. So in that way, it is also the opposite of gravity which is stronger when objects are closer together.

Since no one knows much about dark energy, let us use the rest of this worksheet to list and describe what we do know about energy. You might talk about energy and how it is made from matter using the sun as an example. You might discuss waves and their properties. Perhaps you'd like to list the forms of energy that are electromagnetic radiation. What do you know about energy?

Name _____

Name _____

Benefits of The Moon

13. He causes the night to enter in upon the day, and He causes the day to enter in upon the night, and He has made subservient (to you) the sun and the moon; each one follows its course to an appointed time; this is Allah, your Lord, His is the kingdom; and those whom you call upon besides Him do not control a straw. (35:13)

Check off the benefits that we may get from the moon:

- ☐ The moon causes the high tide/ low tide of the ocean
- ☐ The moon's gravity keeps our tilt stable as we orbit, so we have 4 seasons.
- ☐ The moon's gravity also keeps us from spinning too fast
- ☐ Animals use the moon to navigate, to know when to spawn, to hunt or to hide

Explain one of the checked benefits, or add a benefit that is not listed:

Name _____

Name _____

Magnetism: Spin Up and Down

36. Glory be to Him Who created pairs of all things, of what the earth grows, and of their kind and of what they do not know. [Ya Sin,36]

When Allah made the world, he made everything in pairs. Some of these pairs are opposites that will cancel each other out. Matter had anti matter. Energy was paired with dark energy. Other pairs are about balance. Within the atom, the positive charge had the negative charge to balance it. In this miraculous way, He made even the electrons inside the atom organized into pairs as well.

(draw diagram of atom as provided by your teacher here)

How do they match each other? Electrons are always in motion. They are the negative charges that are orbiting the nucleus. But their orbitals are at a certain distance. Much like the planets stay in their particular orbits around the sun. But in the case of electrons, you can have 2 to 32 electrons in an orbit (known as a shell). When there is an even number, all of them are "paired" by their direction of spin. In each pair, one electron spins clockwise, and the other counterclockwise. This is because the magnetic field of one is pointing upwards, and the other is pointing downwards. In this way, the electrons are not going to repel each other out of the orbit. If they weren't paired, they'd push each other out!

Name _____

Name _____

Creation of Solar System, Part 2

*38. And We did not create the heavens and the earth
and what is between them in sport.
39. We did not create them both but with the truth,
but most of them do not know.*

1. Look for the images of an accretion disk, which is a new solar system being formed (go to http://www.earlyearthcentral.com/star_page.html)

2. Draw the process of a new solar system forming.

3. Investigate the "Big Bang". How did the famous scientist Stephen Hawking came up with the idea that something came from nothing?

Name _____

Name _____

Index

Allah
    What the creation says about Him, 23
Arabic
    Periods (yaum), 7
    subservient, 39
Comets
    composition, compared to Asteroids, 41
Day, 19
    Arabic for, 7
Energy, 43
Life. *See* Water
Light, 11
    Length of Day, 19
Magnetism
    in electron pairs, 47
    Magnetosphere, 35
Meteors, 13
Moon
    Benefits, 45
    Phases, 17

Night
    Length of, 19
    Metaphor, 27
Periods of creation, 7, 9
Solar System
    Accretion disk, 49
    Diagram, 15
Stars
    Astrolabe, 33
    Astrology, 21
    Benefit of, 21
    Creation of, 3
    Zodiac constellations, 25
Structure of Universe
    Atmosphere protects, 35
    Dark Energy, 43
    Dark Matter, 29, 31
    Magnetosphere, 35
Sun
    Serves us (benefits of), 37
    Sundial, 37
Water, 5

Name _____

www.ingramcontent.com/pod-product-compliance
Lightning Source LLC
LaVergne TN
LVHW081400060426
835510LV00016B/1918